ALL ABOUT OWLS

Jim Arnosky

Scholastic Inc.
New York Toronto
London Auckland Sydney

Turn the next two pages
to see a life-size portrait
of a Great-horned owl.

For Kristin and Kelly

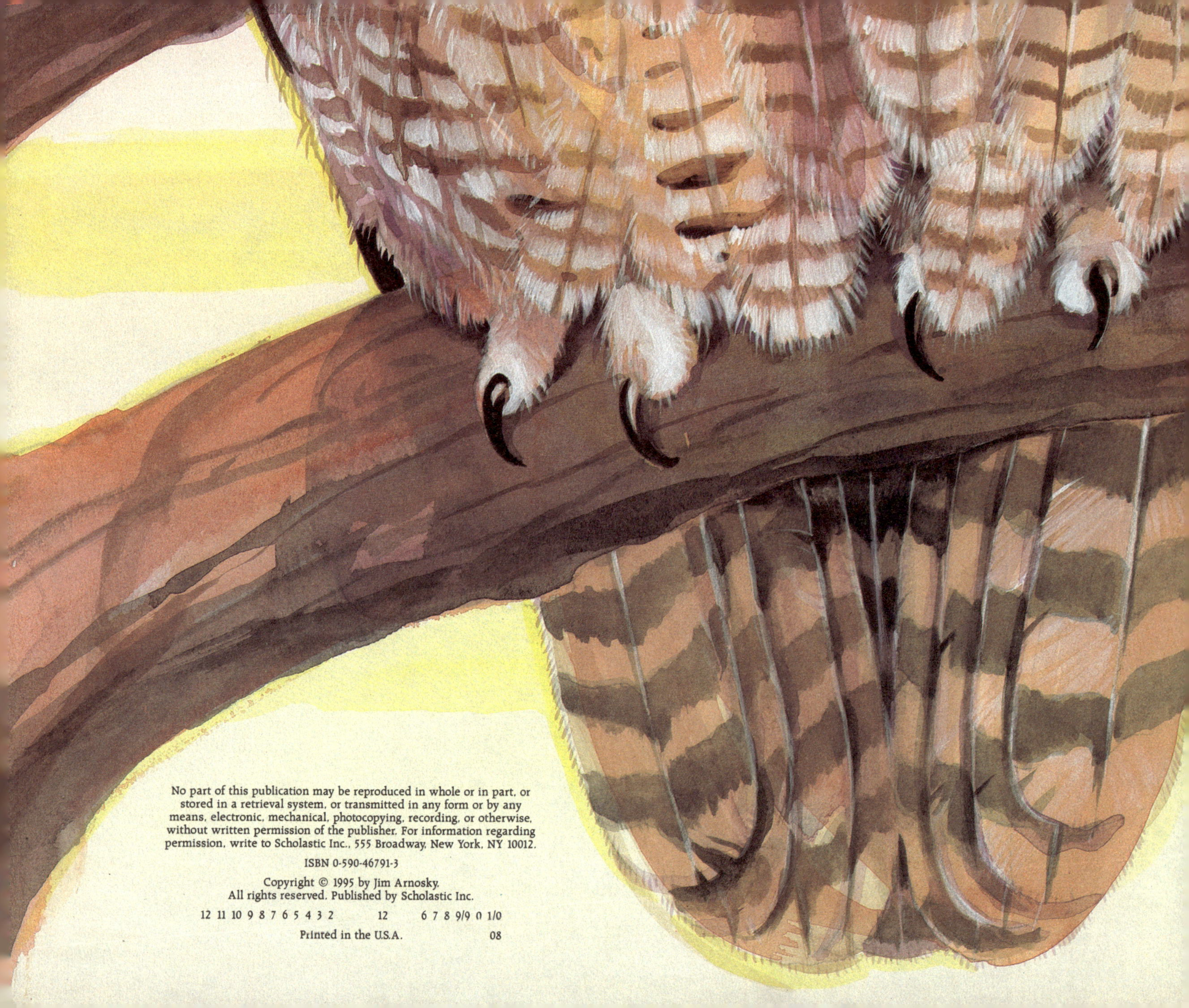

ISBN 0-590-46791-3

12 11 10 9 8 7 6 5 4 3 2 12 6 7 8 9/9 0 1/0

Printed in the U.S.A. 08

Have you ever wondered about owls?
How many kinds of owls are there?
Where do they live?
Why are owls' eyes so big?
What do owls eat?

This book answers all
these questions and more.
It is all about owls!

Great-horned Owl

Owls are birds of prey, which means they hunt, kill, and eat small animals. Eagles and hawks are also birds of prey. But they are daytime birds. Owls are creatures of the night. After sleeping all day, they awaken at dusk and hunt until dawn.

There are 134 species of owls in the world. These 12 show just how different species of owls are in size and color. Owls live in all kinds of places, from the desert to the Arctic tundra.

These are North American owls. Each owl's name, average length, and habitat is listed near its picture.

Screech Owl ▶
10″
Deciduous forests
◀ Boreal Owl
9″–12″
Coniferous
forests
Long-eared Owl ▶
15″
Coniferous forests
Barred Owl ▶
20″
Deciduous forests
Barn Owl ▲
18″
Farms, fields, parks,
old buildings
Great
Gray Owl ▶
24″–33″
Northern forests
(largest of all owls)
▲ Great-horned Owl
18″–25″
Coniferous and deciduous forests

All owls, big or small, have large, rounded heads with flat facial discs, sharply hooked beaks, and big eyes that face forward. Here is what a Saw-whet Owl looks like close up.

Although owls appear to be plump and heavy, they weigh very little. Most of what you see are loose, fluffy feathers. Underneath its feathers, this Saw-whet Owl's body is about the same size as a robin's body!

Saw-whet Owl
(shown actual size)

An owl's facial discs aid its night vision by reflecting all available light into the owl's eyes.

All the feathers on an owl, even the stiffer wing feathers, have soft edges so that the owl can fly silently.

Owls have two toes forward and two toes backward. One rear toe on each foot can pivot forward.

The so-called "eared" owls or "horned" owls are named for the tufts of feathers on their heads that look like ears or horns.

Owls' eyes are large with dark pupils. In the darkness, the pupils can widen to almost the entire size of the eye to gather all available light.

Great-horned Owl

An owl's actual ears are hidden inside a fold of skin on each side of the owl's head.

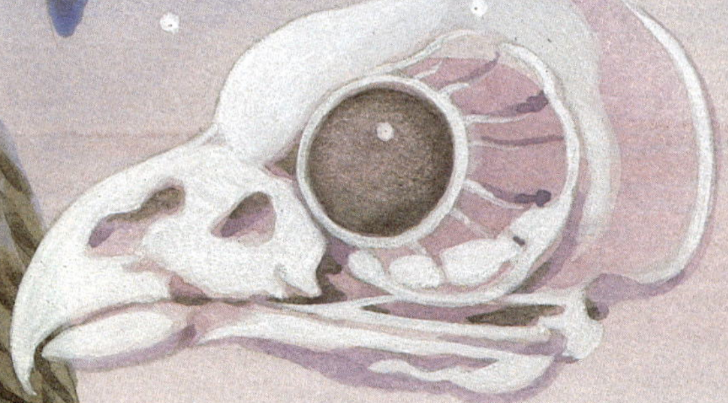

In an owl's skull, the eye sockets hold the eyeballs firmly facing forward, giving owls a wider range of binocular vision than all other birds. Binocular vision enables the owl to judge the distance of its prey.

Owls' eyes are fixed in their sockets and cannot move very much. To look from side to side, an owl must turn its head.

Like all birds, owls have a top, a bottom, and a third eyelid. The third eyelid is transparent and is only for cleaning and moistening the eyeball.

An owl can swivel its head around to look backward.

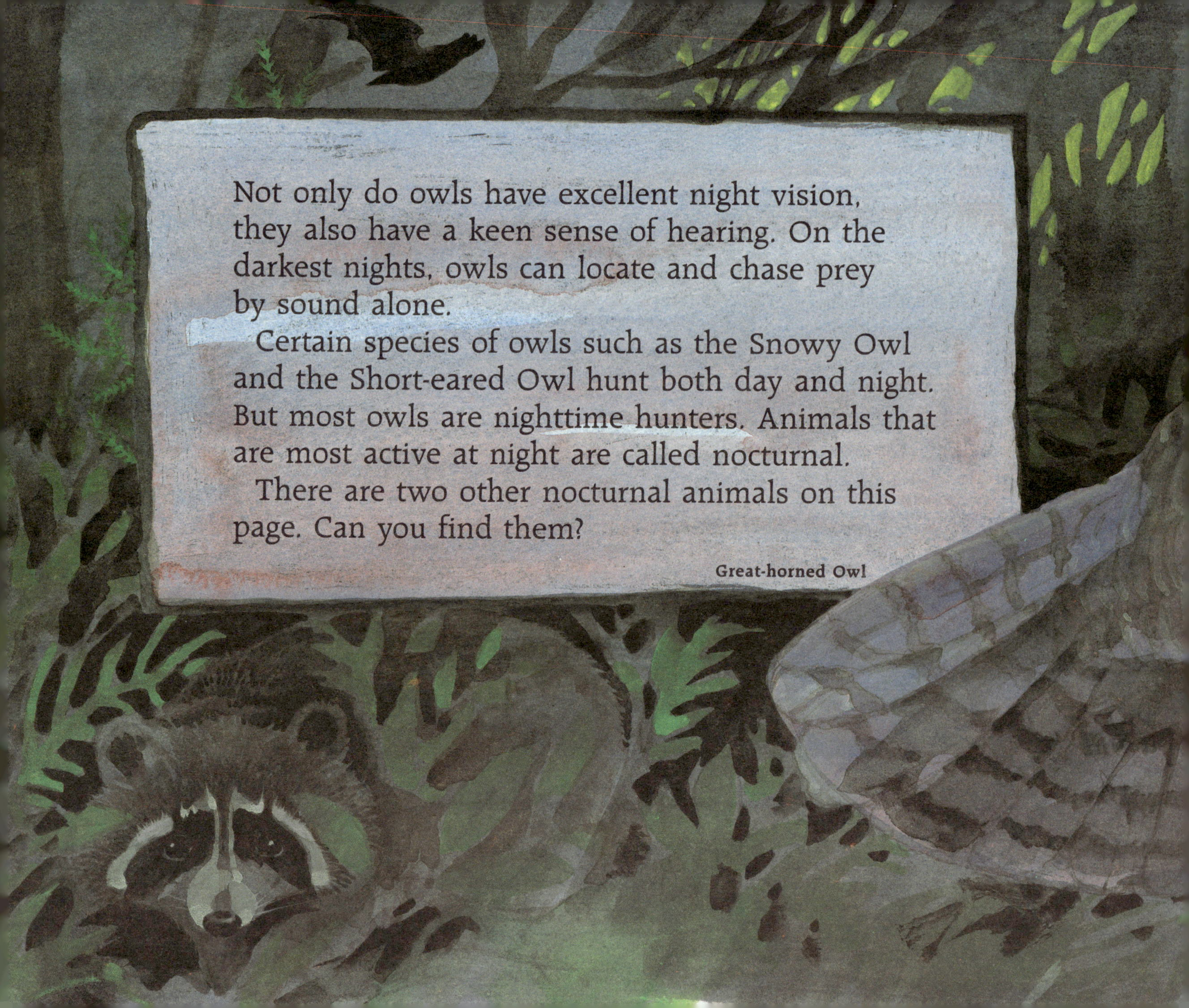

Not only do owls have excellent night vision, they also have a keen sense of hearing. On the darkest nights, owls can locate and chase prey by sound alone.

Certain species of owls such as the Snowy Owl and the Short-eared Owl hunt both day and night. But most owls are nighttime hunters. Animals that are most active at night are called nocturnal.

There are two other nocturnal animals on this page. Can you find them?

Great-horned Owl

Because owls are nocturnal, people rarely see them. The easiest way to tell which owls are living around you is to listen for their calls.

Call of the Barred Owl

HOO-WHOOOO

Call of the Great-horned Owl

HOO-HOO-HOO-WHO

There are as many different owl calls as there are species of owls. But the three most commonly heard and the easiest to identify are the calls of the Great-horned Owl, the Barred Owl, and the Screech Owl.

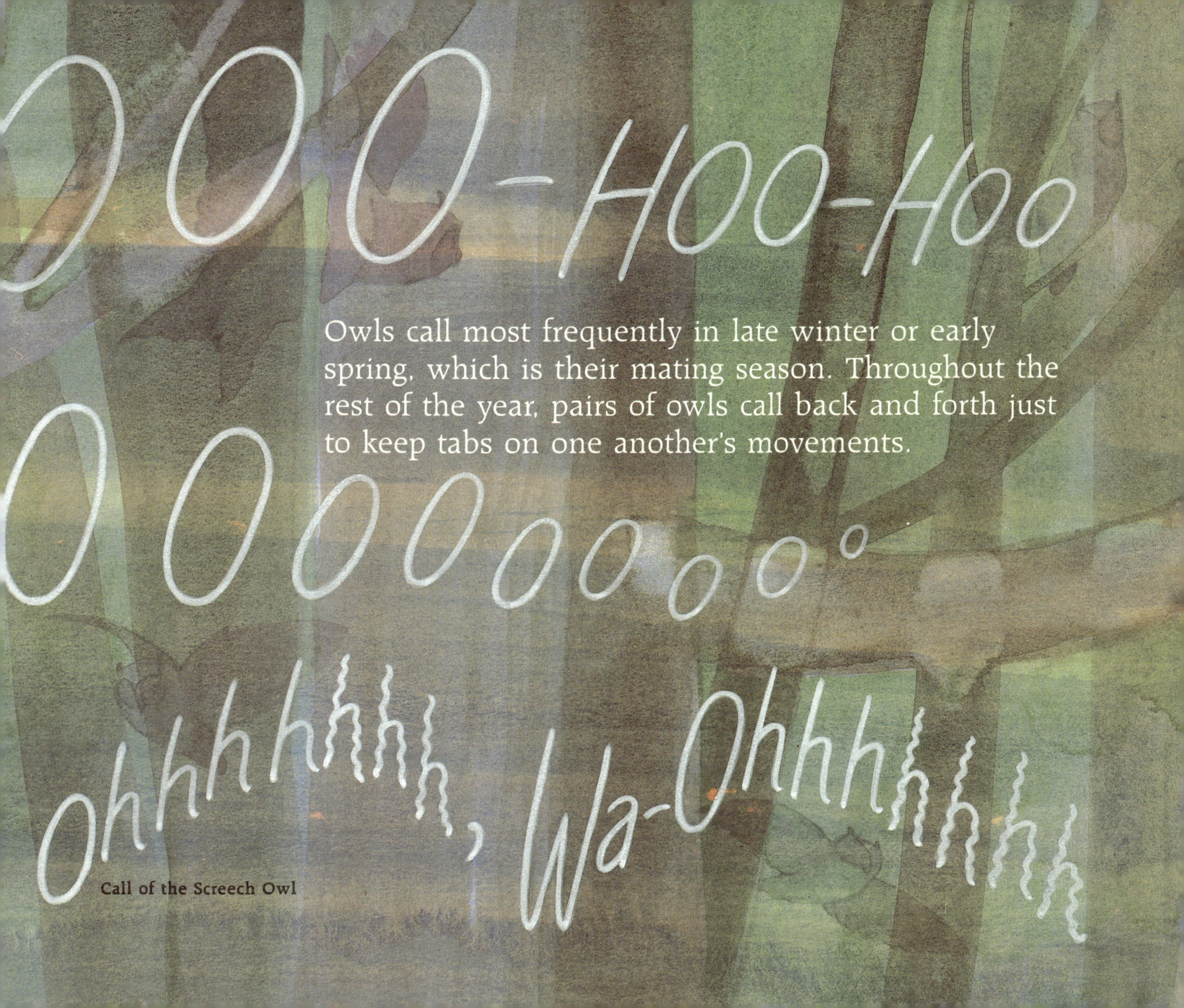

OOO-HOO-HOO

Owls call most frequently in late winter or early spring, which is their mating season. Throughout the rest of the year, pairs of owls call back and forth just to keep tabs on one another's movements.

Oooooooooo

Ohhhhhhhh, Wa-Ohhhhhhhh

Call of the Screech Owl

All owl eggs are white and almost round. The size of an owl egg depends on the size of the owl. Depending on the species, owls lay from two to seven eggs.

Actual size of a Barred Owl's egg with a peek at the hatchling inside

Snowy Owls and Short-eared Owls nest on the ground. Burrowing Owls nest under the ground. All other species of owls nest high up in the natural cavities of trees, abandoned woodpecker holes, or deserted crow or squirrel nests.

Newborn owls are covered with soft, downy feathers and cannot fly. At first they eat only food that is regurgitated, or spit up, by their parents. Very soon, the owlets are fed torn bits of prey. At six weeks of age, owls are big enough to fly and leave the nest and learn to hunt for themselves.

Barred Owl owlets in
tree cavity nest

Saw-whet Owl

When hunting, owls watch for any movement in the night. They listen for the rustling sounds of mice on the ground or the wafting sound of moth wings in the air.

Owls eat more rodents than anything else they eat. The smaller species of owls will eat insects, especially moths.

Larger owls also eat weasels, rabbits, snakes, and other birds, including smaller owls.

An owl uses its beak to pick apart large prey. Small animals are swallowed whole.

Indigestible bones and fur are coughed up in the form of a compact wad called a pellet.

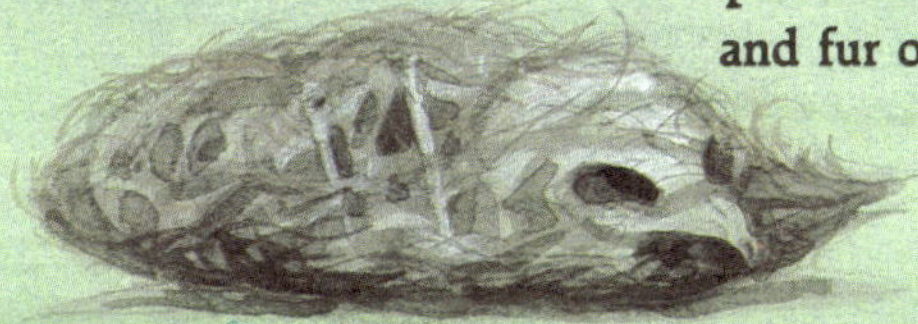

Actual size of a Saw-whet Owl pellet containing bones and fur of a mouse

An owl is a crow's worst enemy. Night-hunting owls pounce on sleeping crows. Owls raid the nests of crows and other birds. If an owl is discovered by a crow during daylight hours, the entire flock of crows will mob the owl, repeatedly diving at its head to drive it away. So during the day, owls always try to stay out of sight.

Great-horned Owl

Barn Owl

In North America it is against the law for any person to harass or harm owls. Owls are just as beneficial in cities as they are in the countryside, because they feed mainly on rodents. This helps keep the rodent population in control.

And, of course, wherever they live, owls add a sense of mystery to the sounds and silence of the night.

Michelle Sherburne

JIM ARNOSKY is the award-winning illustrator of more than 30 books for children, and both author and illustrator of 35 more. He lives with his family in northern Vermont.

For *All About Owls*, Jim Arnosky tried to answer all his own questions about owls. He went out at night to listen for owls. He went into the dark woods to photograph owls. And he sketched, painted, and video-taped owls.

This is the second book in Jim Arnosky's *All About . . .* series for Scholastic. The first book is *All About Alligators.* The third book is *All About Deer.* There will be six more.